Orcas

Orcas

Melissa Gish

Living Wild

CREATIVE EDUCATION
CREATIVE PAPERBACKS

Published by Creative Education and Creative Paperbacks
P.O. Box 227, Mankato, Minnesota 56002
Creative Education and Creative Paperbacks are imprints
of The Creative Company
www.thecreativecompany.us

Book design by Tom Morgan (www.bluedes.com)
Art direction by Rita Marshall
Book production by Alison Derry
Edited by Jill Kalz

Photographs by Adobe Stock (OlgaChernyak), Alamy (Krisztina Balotay, INTERFOTO, Ivan Kotliar, Mary Evans Picture Library, Old Books Images, Blaize Pascall, Science History Images, Volvox Inc), Corbis (Werner Forman), Dreamstime (C-foto, Ferdericb, Uniqueglen), Flickr (Biodiversity Heritage Library), Free Vintage Illustrations (Public Domain), Getty (Cris Bouronclie/AFP), iStock (Frederic Borsset, Karoline Cullen, duncan1890, Richard Goerg, Tatiana Ivokovich, Benjamin Jessop, Hakan Karlsson), Pexels (Messina), Shutterstock (Martens Tom)

Library of Congress Cataloging-in-Publication Data
Names: Gish, Melissa, author.
Title: Orcas / by Melissa Gish.
Description: Mankato, Minnesota : Creative Education/Creative Paperbacks, [2024] | Series: Living wild | Includes bibliographical references and index. | Audience: Ages 10–14 | Audience: Grades 7–9 | Summary: "A look at orcas, including their habitats, physical characteristics such as their unique coloration, behaviors, relationships with humans, and protected status in the world today"— Provided by publisher.
Identifiers: LCCN 2022022640 (print) | LCCN 2022022641 (ebook) | ISBN 9781640266025 (library binding) | ISBN 9781682771570 (paperback) | ISBN 9781640007215 (ebook)
Subjects: LCSH: Killer whale—Juvenile literature.
Classification: LCC QL737.C432 G575 2024 (print) | LCC QL737.C432 (ebook) | DDC 599.53/6—dc23/eng/20220520
LC record available at https://lccn.loc.gov/2022022640
LC ebook record available at https://lccn.loc.gov/2022022641

Printed in China

CONTENTS

A group of orcas pauses its day's travels to gather in the chilly water of Johnstone Strait, between Canada's Vancouver Island and British Columbia. The sun shines brightly, but it doesn't warm the water above its early autumn average of 43 degrees Fahrenheit (6 degrees Celsius). Some of the orcas flop over onto their backs, gently rubbing against each other and soaking up the sunshine. Two youngsters, each weighing more than 1,500 pounds (680 kilograms), rush to a big male who is floating on his back. The youngsters let out loud squeaks, bumping the male on the face and then dashing away. He rolls onto his belly and floats patiently, allowing the youngsters to bump, rub, and flop on him. It is playtime for the young orcas, and the adults—brothers and sisters, aunts and uncles—all relax and allow the lively youth to enjoy the calm afternoon.

An orca has 50 to 54 bones in its back but none in its flukes, the two flat pads at the end of the tail.

Whale Killers

The orca is one of Earth's top predators, hunting and eating just about anything it can catch in all five of the world's oceans. Groups of orcas will even attack the largest animals in the sea—whales. Orcas are also called killer whales, derived from the name *ballena asesina*, or "whale killer," which was what 18th-century sailors called them when they saw groups of orcas ripping apart enormous whales.

Orcas' scientific name, *Orcinus orca*, relates to Orcus, a god of the underworld in Roman **mythology**. In Latin, *orcinus* means "from hell," and *orc* means "whale."

While orcas are found all around the world, they are most common in the Arctic and Southern Oceans. They also live off the icy Pacific coasts of Alaska and Canada, the frigid Atlantic coasts of Iceland and Norway, and in the chilly waters of the

Southern Ocean near Argentina. Because most orcas are constantly on the move in search of food, they may venture into warm waters, such as those of the Indian Ocean and Gulf of Mexico, and up rivers such as Oregon's Columbia River.

The orca is not a whale but rather the largest member of the dolphin family Delphinidae. The orca's closest relatives are the pygmy killer whale and the other 31 species of dolphins. Orcas are marine mammals, meaning they live in the water and belong to a class of animals that, except for the egg-laying platypus and echidna, give birth to live young and produce milk to feed them.

Like all mammals, orcas are warm-blooded. This means that their bodies maintain a constant temperature that is usually warmer than their surroundings. To help stay warm, the orca has 3 to 4 inches (7.6–10 centimeters) of thick fat, called blubber, just beneath its skin. Blood vessels in the flukes, flippers, and dorsal fin (which is located on the animal's back) also help the orca control its temperature.

Since mammals need to breathe air, orcas must regularly swim to the surface of the water. An orca breathes through its blowhole, a type of nostril located on top of its head. Once it reaches the surface, the orca opens its blowhole, sucks in a great amount of air, and then pinches its blowhole shut as it dives into the water.

Orcas have enormous lungs that allow them to stay underwater for up to 15 minutes. Such lengthy dives are rare, however. Usually, orcas remain underwater for no more than five minutes. Most of the time, they swim near the water's surface to take a breath every 30 seconds.

Toothed whales such as the orca have one blowhole, while baleen whales such as the humpback have two.

Where in the World They Live

The single species of orca is typically found in colder ocean waters around the globe. It primarily frequents the Arctic, North Atlantic, and North Pacific Oceans in the Northern Hemisphere and the Southern Ocean in the Southern Hemisphere, and it is occasionally seen in warmer gulfs and other waters as well. The numbers on the map represent some common locations.

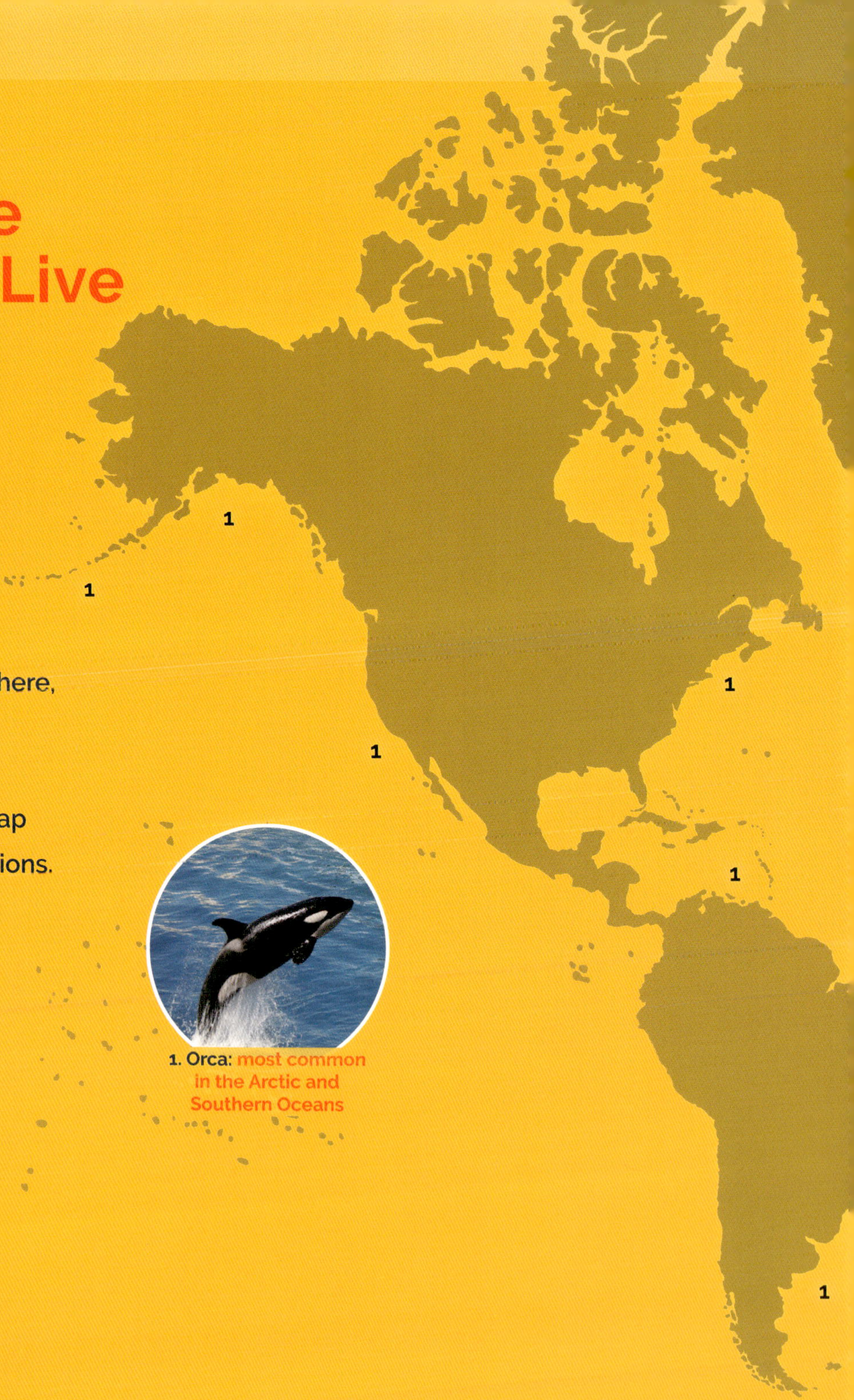

1
1
1
1
1
1

1. Orca: most common in the Arctic and Southern Oceans

1
1
1
1
1
1
1

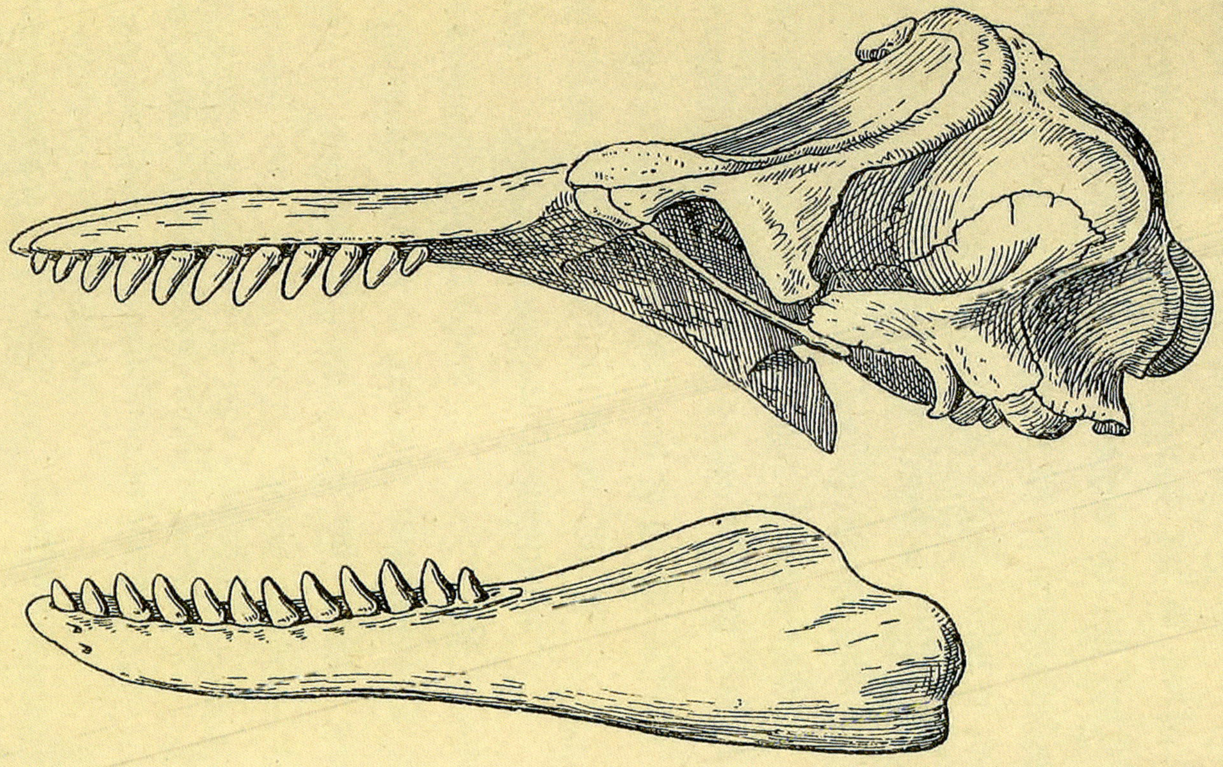

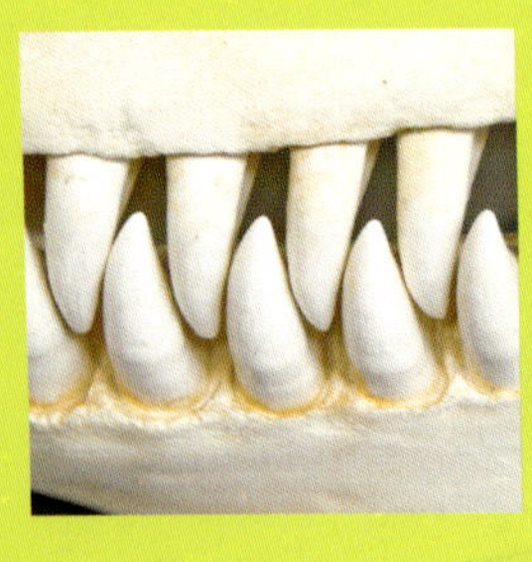

Each orca tooth, which curves inward to tightly grip prey, measures about 3 inches (7.6 cm) long and 1 inch (2.5 cm) thick.

Despite being related to orcas, dolphins instinctively flee when the predators come too near.

Orcas are the only species of dolphin with black skin. They have distinctive white eye and belly markings that are unique to each orca and are used by researchers to distinguish one orca from another. Individuals can also be recognized by the gray or whitish mark behind the dorsal fin, called a saddle patch, which varies in shape and coloration from one orca to another.

Orcas are massive creatures, nearly the length of a school bus. Females typically weigh about 8,000 pounds (3,629 kg) and grow up to 22 feet (6.7 meters) in length. The average male weighs up to 12,000 pounds (5,443 kg) and is more than 26 feet (7.9 m) long. Adult orcas have more than 40 interlocking teeth. The snout is tapered, the body is smooth and sleek, and the tail is slender. This streamlined shape allows for little **resistance** as an orca cuts through the water, enabling it to move at great speeds while chasing prey and to dive and surface quickly.

The tail gives an orca its speed. At the end of the tail are two flat, boneless pads called flukes. The orca uses the muscles in its back and tail to wave the flukes up and down. This motion propels the orca forward—at speeds of up to 30 miles (48 kilometers) per hour.

Orcas have the largest brain of all dolphins, with a skull measuring more than 36 inches (91 cm) long.

To help steer as it swims, the orca uses its pectoral flippers. These are located on each side of the chest below and behind the head. The flippers have bones like those in a human's hand, enabling the orca to twist its pectoral flippers somewhat sideways in order to slow down or stop swimming. The orca uses its dorsal fin to steady itself as it swims. Without the weight of this boneless appendage, an orca would roll upside down in the water. A young male's or female's dorsal fin is usually 3 to 4 feet (0.9–1.2 m) tall and curves to the right or left. When a male reaches 12 to 14 years of age, his dorsal fin grows taller and straightens, eventually reaching up to 6 feet (1.8 m) in height.

Orcas communicate vocally with each other by making clicks, whistles, pulsed calls, and pops. Whistles change pitch, like songs, and are used to communicate socially in groups. Pulsed calls sound like squeaks or screeches, while pops are low-frequency sounds. Pulsed calls and clicks are also used during **echolocation**.

Echolocation helps orcas to "see" underwater. To echolocate, an orca sends out pulses of sound from various nasal sacs and cavities. These sounds hit objects in their path and bounce back, like echoes, and are captured by a fatty organ in the forehead called a melon. The orca then "reads" the echoes, determining the location, shape, and size of the objects around it—including prey. The objects can be as big as a whale or as small as a fish. With this ability, orcas can hunt even in total darkness. Using echolocation, orcas can determine the differences between humans and prey animals such as seals. This is the reason, researchers believe, that orcas do not attack humans in the ocean.

To dine on a school of fish, orcas first smack them with their tails to knock them out or blow bubbles to force them to the surface.

Orcas living off the coast of Norway feed on schools of herring, and orcas in New Zealand's waters eat sharks and stingrays.

When a pod of orcas gathers together and floats briefly on the water's surface, the action is called logging.

Family Matters

Orcas live in family groups called pods, which are led by a dominant female known as the matriarch. A pod is made up of several mothers and their offspring and includes 10 to 20 members that travel together for their entire lives.

By biting, head-butting, and scraping her teeth on other orcas' skin, a matriarch shows her pod members that she is in charge. When she dies, one of her daughters will take over.

There are three distinct kinds of orca pods: offshore, **transient**, and resident. Offshore orcas always stay far from land. Little is known about their behavior, as their pods are very small, and they are not easy to find. Transients tend to hunt silently in small groups and sneak up on their prey, including seals, sea lions, walruses, and dolphins. Individual transients may move from pod to pod throughout their lives. Offshore and transient orcas **migrate** in summer and winter to follow stocks of fish that also migrate. Residents tend to stay in the same pods their entire lives. They do not migrate over long distances but instead travel around a specific range. They eat smaller prey, passing up seals in favor

Orcas and polar bears are the only natural predators of 4,000-pound (1,814 kg) walruses.

of fish, and they spend lots of time vocally communicating with members of their pod.

Body language is another important form of communication. When an orca propels itself out of the water quickly and falls back down with a big splash, it is breaching. This is one way the animal shows off its strength. To show dominance, an orca might loudly smack its tail or paddle its flippers on the water's surface. In addition to such displays, orcas utilize many other kinds of signals to tell each other where food is located or where danger is lurking.

Pod members stay close together. Many pods that live near each other can be classified as a clan. The orcas of a clan are all related to one another. Because orcas will not mate with relatives, several clans will gather in a community of 100 or more individuals that are not all related to facilitate mating. Resident communities will mate and then separate, but clans of transient and offshore communities, also known as superpods, may travel and hunt together in addition to mating.

Mating occurs only during specific seasons, which vary depending on the climate in which the pod lives. Summer is the most common time for orca mating to take place. Female orcas mature when they are about 15 feet (4.6 m) long, which could happen any time between the ages of 6 and 10 years. A female will breed once every 3 to 10 years until she is 40 years old. Males mature when they are about 20 feet (6.1 m) long and their dorsal fins have reached full height. This happens when the males are about 12 to 14 years old.

The first thing that a female notices about a potential mate is the dorsal fin. Females are usually most attracted to older males whose dorsal fins are very tall. When a female decides to pay attention to a male, she will watch him slap the surface of the water with his flippers and flukes. Then the male will chase the female around, splashing and bumping her body with his head and body. He will swim close to her, rubbing his skin against hers. If she likes him, she will invite him to mate with her.

Orcas mate while swimming together. A female will mate with several males to increase her chances of producing strong offspring. After mating, the community breaks up into smaller clans or pods. Because the fathers move back to their own pods, they take no part in the care of their offspring, called calves. Males in a calf's pod—its brothers and uncles—will protect and teach it throughout its early years.

The female carries her single calf for 15 to 17 months; twins are rare. When the calf is ready to be born, the mother swims in a twisting, rolling motion to help push out the calf, which emerges tail first. Instinct drives the newborn calf toward the water's surface. The mother lifts the calf with her snout to help it take its first breath.

Newborn calves are 7 to 9 feet (2.1–2.7 m) long. They can weigh up to 400 pounds (181 kg). For the first few days of a calf's life, its flippers and flukes are floppy. And for the first year, its light-colored areas of skin are creamy instead of white.

Over her lifetime, a healthy female orca may give birth to five or six calves.

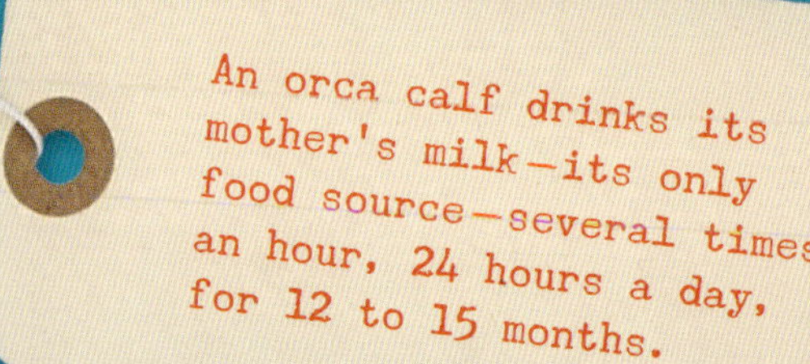

The milk produced by a female orca is full of the fat and nutrients that her calf needs to grow. The calf presses its mouth against its mother's body, and she squirts milk directly into its mouth. At about 12 weeks, the calf's upper teeth emerge, and about 4 weeks after that, the lower teeth begin to grow. The calf then begins to eat prey with other members of its pod, but it also continues to rely on its mother's milk until it is one year old. By that time, the calf has gained about 1,000 pounds (454 kg).

An orca calf's life is filled with danger. It is vulnerable to attack by predators such as sharks, many of which can be twice as long as an orca calf. About half of all young orcas die within the first year. Many perish at birth because they are not big enough or strong enough to survive in the sea. Predators, disease, and other factors claim the rest.

Orcas that are related to one another help care for calves, teaching them how to hunt, recognize danger, and communicate. Young orcas learn vocal communication and body language by mimicking the older orcas in the pod. Each pod has its own vocal language variation, or dialect, that is passed down from the older orcas to the calves.

The orca has a sleek, streamlined body that is tapered at both ends, a shape described as fusiform.

Star of the Show

Today, people understand that orcas are highly intelligent and trainable, but this was not always the case. The ancient Greeks viewed orcas as monsters bent on bringing bad luck to sailors. Common dolphins represented good luck, and seeing orcas attack and devour dolphins struck fear and superstition into the hearts of Greek sailors. Their experiences were the basis of myths and legends surrounding orcas.

Orcas are kept in captivity around the world, from France's Marineland (*pictured*) to Venezuela's Acuario Mundo Marino.

The first written account of orcas was published in *Natural History*, an encyclopedia written by Pliny the Elder, an ancient Roman naturalist and author, in A.D. 77. He noted the animals' "savage teeth" and described their attacks on great whales with battle imagery, saying that orcas "charge and pierce [the great whales] like warships ramming." Such stories of monstrous "orcs" persisted for many centuries, from Italy to Sweden. In his 1555 *History of the Northern Peoples*, Swedish writer Olaus Magnus wrote how orcas used their "ferocious teeth . . . as brigantines [fast sailing ships] do their prows [the pointed front parts of ships]" to rip into the body of a great whale.

In the 19th and 20th centuries, when scientists began traveling with sailors and whalers on their voyages, the accounts of orcas became more scientifically accurate. People began to understand that orcas were intelligent animals, and organizations in Canada, the United States, and some northern European nations established research projects aimed at learning more about orcas.

But just as the first observations of orca intelligence were being published, the whaling industry began to see the financial value of capturing orcas. By the early 1900s, stocks of regularly hunted whales, such as fin whales and right whales, had declined drastically due to overhunting. Because orcas were abundant, whalers turned their attention to these creatures instead. Almost 10,000 were killed before whaling ceased in most countries in the 1980s, and orcas were spared from having to face extinction.

Orcas are still subsistence-hunted by certain native peoples of the Arctic. For thousands of years, these cultures have valued the orca not only as a food source but also as a cultural icon. Some of the cultures that have made the orca a major part of their mythologies include Native Alaskan groups such as the Aleut, Chugach, and Inuit; the Tlingit (*KLING-git*), Kwakwaka'wakw (*KWALK-walk-ya-WALK-wuh*), Haida, and Nootka peoples of British Columbia; and the Yupik of Russia.

Pliny the Elder (*above*) took much of the biological information for his *Natural History* from the Greek thinker Aristotle.

Crumbling wrecks are all that remain of the wooden whaling ships that were used into the early 20th century.

One belief that is shared by many native cultures is that orcas can come ashore and transform into humans, and humans can go underwater and become orcas. In Haida tradition, orcas are believed to have an underwater society like that of humans. Likewise, Kwakwaka'wakw mythology tells that orcas are commanded by the powerful chief of an underwater kingdom, and they must be respected.

Today, orcas are kept in captivity for research and enjoyment. The first orca to be captured was named Moby Doll—although she was later determined to be a male. In 1964, Moby Doll was shot with a harpoon, which is a large, barbed spear attached to a rope, at Saturna Island in British Columbia and dragged to shore to be used as an artist's model for a sculpture. The orca was kept in a cage and fed fish by his captors in the first recorded experience of an orca accepting food directly from a human's hand. Moby Doll died 87 days after his capture, but his intelligent behavior sparked people's interest in working closely with orcas.

In 1965, a 10,000-pound (4,536 kg) orca became entangled in a fishing net off the coast of British Columbia and was sold by its captors. Given the name Namu, the orca became the first of its kind to be put on public display and develop a relationship with a human. The human was Ted Griffin, owner of the Seattle Public Aquarium (now known as the Seattle Aquarium) in Washington state. After studying Namu for a month, Griffin determined that he would be safe swimming with the orca. He then became the first person to climb into a pool with an orca, and for the next 11 months, thousands of people visited the aquarium to see Namu and Griffin.

Captive orcas have been known to express behaviors suggesting that they form bonds with their trainers.

Animal Tale: Natcitlaneh Creates the Orca

Tlingit culture is tied to the icy ocean and its creatures. The native people of Canada's Pacific coast believe that all things were created for a reason and that people and animals share a special relationship based on respect. The following Tlingit story explains how the orca, called "Teek" in the Tlingit language, came to be and why people should respect rather than fear it.

Long ago, there was a skilled sea lion hunter named Natcitlaneh. He met and fell in love with the daughter of a neighboring chief, who gave Natcitlaneh permission to marry her. Natcitlaneh moved to the chief's village and married his daughter. He proved his worth by hunting many sea lions and providing a bounty for the village. This made his wife's brothers jealous.

One day, the brothers invited Natcitlaneh to join them in hunting sea lions on the ice floes far from shore. They all climbed into their canoes and headed out to sea. When Natcitlaneh could no longer see the shore, the brothers tipped his canoe and sent him tumbling into the icy water. Then they took his canoe and left him floundering in the sea.

The sea lions, respecting his skill as a hunter, rescued Natcitlaneh and took him to their village in the ice caves of a glacier. The sea lions' chief was sick, but Natcitlaneh healed him. This pleased the sea lions, who then gave Natcitlaneh a magical knife. With the knife, they told him, he could carve a magnificent new animal that would carry him home.

The sea lions took Natcitlaneh to a dense forest on a nearby island, where he cut down a yellow cedar tree and began to carve with his magical knife. He carved a great fishlike creature with broad flippers, a tall dorsal fin, a strong tail, and mighty jaws. When he placed the carving in the sea, it came to life.

Natcitlaneh climbed on the orca's back and rode it across the sea toward his wife and their village. As they traveled, Natcitlaneh told the orca the story of how his brothers-in-law had betrayed him and left him to die.

As they approached the shore, Natcitlaneh saw his brothers-in-law in their canoes. They were going to hunt sea lions. Before Natcitlaneh could stop the orca, the great

beast swam to the canoes and smashed them with its tail. The brothers-in-law fell into the water, thrashing and crying out. The orca was so angry at the men's betrayal of Natcitlaneh that it leaped on them, crushing their canoes and drowning them.

Natcitlaneh was avenged, but he felt regret for allowing his wife's brothers to be killed. He made the orca promise that it would never again harm a human being. The orca promised, and then it returned to the sea. It swam in circles back and forth along the shore, ever watchful over Natcitlaneh, but it never hurt anyone again.

The people of the village were amazed by Natcitlaneh's story, and when they saw for themselves that the orca had kept its promise, they took the honorable orca as their village symbol. And to this day, orcas never, ever eat humans.

Rulers of the Sea

Millions of years ago, orca ancestors lived on land. The orca likely **evolved** from four-legged mammals that lived in and around streams and shallow lakes and ate plants.

One orca ancestor was the cat-sized *Indohyus*, which lived about 50 million years ago and looked like a miniature deer. It had a thick hide like modern wading animals, such as the hippopotamus, and dense fur like that of a beaver.

As these prehistoric mammals began to spend more time in the water, they lost their legs, instead gaining webbed limbs and eventually flippers. Over many millions of years, these prehistoric creatures grew very large, outcompeting any other predators. They gradually became more torpedo-shaped and developed a dorsal fin. Fossil research indicates that primitive orcas began to develop the ability to echolocate 10 million years ago.

The use of echolocation to hunt prey is the subject of much current orca research. The Northwest Fisheries Science Center is using digital tags (DTAG) to record orca sounds in hopes of

When an orca pops straight out of the water and then silently slips back down, it is spyhopping—checking out the surface for prey.

Wind erosion in Egypt has uncovered a treasure trove of fossilized marine mammals at least 40 million years old.

learning how excess noise from boats affects the animals' ability to hunt. Orcas use slow clicks at the surface of the water to search for prey. The slow clicks are subtle enough so fish can't hear them. When an orca detects a fish, it dives down, increasing the rate of its clicks until they sound like a constant buzz as the orca closes in on its prey. This noise is easier for fish to hear and sometimes the prey is fast enough to get away. However, underwater noise from nearby boats makes it harder for orcas to hear the echoes of their clicks. They must swim closer to their prey, which increases the chance of the fish hearing them.

Another echolocation project is studying a pod of resident orcas that hunt along the Pacific coast. They use echolocation to find their preferred prey, Chinook salmon. The orca's echolocation is so precise that it can tell one species of fish from another and pinpoint the oldest, largest salmon. Understanding how orcas hunt with echolocation can help researchers better protect the mammals from noise disturbances that can interrupt their feeding habits. Slower boat speeds help. They don't produce as much noise. In the United States, boaters must stay at least 300 yards (274 m) from orcas and travel at speeds less than 8 miles (13 km) per hour when orcas are spotted within a half mile (805 m).

Researchers use tracking devices to monitor orca populations. Information on the orcas and their favorite source of prey can be collected using photo-identification research—taking photographs of surfacing orcas and keeping records of the sightings—and by **satellite** tracking. Tags are attached to the orcas' dorsal fins or flukes. These

Whale watching is a popular activity, with millions of people across more than 100 countries taking to the sea annually.

tags have **Global Positioning System** (GPS) tracking devices on them. The information gathered from the GPS devices helps researchers count pod and clan populations, understand how and where the orcas travel during the seasons, and learn about orca diving and hunting behaviors. Resident pods are the easiest to monitor, as they typically stay in the same area. Transient pods are more difficult to study, as their range is much greater. Offshore pods are generally not studied, since they are nearly impossible to find and track.

Many countries have launched research projects specifically devoted to orcas. The Far East Russia Orca Project is the only orca research project in Russian waters. Scientists study the animals off the Kamchatka Peninsula in the northern Pacific Ocean. The Orca Network, an organization in Washington state, aims to provide information that will lead to the conservation of orcas and their habitats. It sponsors research on the impacts that habitat damage, industrial pollution, and other human activities can have on orca populations.

Another Washington-based research project monitors a pod of resident orcas and has noted a steady decline in the pod's population. While orcas have long been protected from hunting by the Marine Mammal Protection Act of 1972, other human factors, such as pollution from industry and shipping traffic and overfishing of key prey, are known to affect orcas. Additionally, when people expand cities by developing areas near orca habitat, the ocean resources on which orcas depend are affected.

In recent decades, North Atlantic orcas have learned to follow fishing vessels to eat discarded herring.

Researchers around the world have formed networks through which they share information on orcas—generally by means of photo-identification research. The Orca Research Trust, founded by Dr. Ingrid Visser, is an organization that gathers information from researchers in the South Pacific. Dr. Visser's Antarctic Killer Whale Identification Catalogue is a collaborative project that gathers and shares photos of individual orcas in the Southern Ocean.

Scientists at such places as the North Gulf Oceanic Society in Alaska and the Vancouver Aquarium Marine Science Centre in British Columbia rely on photo-identification research to monitor the health of orca populations. They also take **DNA** samples by shooting small darts into the bodies of orcas, which harmlessly extract tiny samples of skin and blubber. Scientists can then analyze the **genes** of orcas to study how female orcas are able to determine which potential mates are not relatives.

With the tissue samples, scientists can also identify the types and amounts of **contaminants** in the orcas' bodies. Some common contaminants are DDT and PCBs. DDT is a chemical used to kill pests on crops. It has a history of washing into waterways, where it is consumed by animals, negatively affecting them in various ways. PCBs are industrial chemicals that also find their way into Earth's water sources, and when consumed, they cause long-term damage to organs such as the liver and stomach.

Other environmental studies focus on the effect of **global warming** on orcas. As ocean temperatures rise, fewer **plankton** survive, resulting in less food for fish—which means less food for orcas. Environmental

Orcas may breach to get rid of dead skin, as their skin cells grow 290 times faster than those on a human arm.

changes that influence a top ocean predator such as the orca may negatively affect every other life form on the **food chain**.

Orca behavior is also a subject of research. Because orcas are highly social animals with complex communication techniques, they are easily disturbed by unnatural activities. Noise caused by industry, military activities (such as sonar testing and underwater explosions), and tourism can disrupt orcas' communication and hunting behaviors. Loud noises in the water can cause orcas to become confused or separated from their pods, or they may accidentally swim too close to shore and get stuck on the beach.

The types of noises that affect orcas and the sounds that orcas make are studied using cameras and audio recording devices. The information gained from such methods helps researchers learn about the role of vocal communication and echolocation on orca social structure, pod movement, and hunting behaviors. For example, it has been learned that orcas call to each other using sound variations that identify individuals—much like calling one another by name.

Being ocean-dwelling animals, orcas are difficult to study. Enough has been learned about them, though, to dispel the many misconceptions about these brainy social creatures. No longer viewed as monsters of the deep, orcas are now referred to as "sea pandas," fascinating animals with great intelligence and charm. Protecting their habitat—Earth's oceans—is integral to keeping orcas alive and sustaining a balanced food chain in the seas.

Glossary

contaminant – a non-natural substance that has a negative effect upon the environment or animals

DNA – deoxyribonucleic acid; a substance found in every living thing that determines the species and individual characteristics of that thing

echolocation – a system used by some animals to locate and identify objects by emitting high-pitched sounds that reflect off the object and return to the animal's ears or other sensory organs

evolve – to gradually develop into a new form

extinction – the act or process of becoming extinct; coming to an end or dying out

food chain – a system in nature in which living things are dependent on each other for food

gene – the basic physical unit of heredity

Global Positioning System – GPS; a system of satellites, computers, and other electronic devices that work together to determine the location of objects or living things that carry a trackable device

global warming – the gradual increase in Earth's temperature that causes changes in climates, or long-term weather conditions, around the world

migrate – to travel from one region or climate to another for feeding or breeding purposes

mythology – a collection of myths, or popular, traditional beliefs or stories that explain how something came to be or that are associated with a person or object

plankton – microscopic algae and animals that drift or float in the ocean

resistance – the slowing effect applied by one thing against another

satellite – a mechanical device launched into space; it may be designed to travel around Earth or toward other planets or the sun

subsistence – relating to production of something at a small-scale level, without extra to trade

transient – staying in a place for only a short time

Orcas slide onto shore to grab seals and sea lions, and they can tip floating slabs of ice with their heads to toss penguins into the sea.

Selected Bibliography

Center for Whale Research. "About Orcas." https://www.whaleresearch.com/about-orcas.

Getten, Mary J. *Communicating with Orcas: The Whales' Perspective.* Charlottesville, Va.: Hampton Roads Publishing, 2006.

Hoyt, Erich. *Orca: The Whale Called Killer.* Buffalo, N.Y.; Richmond Hill, Ontario: Firefly Books, 2019.

The MarineBio Conservation Society. "Orcas (Killer Whales)." https://www.marinebio.org/species/orcas-killer-whales/orcinus-orca/.

Morton, Alexandra. *Listening to Whales: What the Orcas Have Taught Us.* New York, N.Y.: Ballantine, 2002.

Orca Network. "Homepage." http://www.orcanetwork.org.

Index